ENERGY SECTOR STANDARD OF THE PEOPLE'S REPUBLIC OF CHINA

中华人民共和国能源行业标准

Code for Design of Rare and Endangered Plants and Old and Notable Trees Protection for Hydropower Projects

水电工程珍稀濒危植物及古树名木保护设计规范

NB/T 10487-2021

Chief Development Department: China Renewable Energy Engineering Institute

Approval Department: National Energy Administration of the People's Republic of China

Implementation Date: July 1, 2021

China Water & Power Press

Beijing 2024

All rights reserved. No part of this publication may be reproduced, stored in a retrieval system, or transmitted in any form or by any means—electronic, mechanical, photocopying, recording or otherwise, without prior written permission of the publisher.

图书在版编目（CIP）数据

水电工程珍稀濒危植物及古树名木保护设计规范：NB/T 10487-2021 = Code for Design of Rare and Endangered Plants and Old and Notable Trees Protection for Hydropower Projects (NB/T 10487-2021)：英文 / 国家能源局发布. -- 北京：中国水利水电出版社, 2024. 5. -- ISBN 978-7-5226-2680-2

I. Q948.52-65

中国国家版本馆CIP数据核字第2024M2T490号

ENERGY SECTOR STANDARD
OF THE PEOPLE'S REPUBLIC OF CHINA

中华人民共和国能源行业标准

Code for Design of Rare and Endangered Plants
and Old and Notable Trees Protection
for Hydropower Projects

水电工程珍稀濒危植物及古树名木保护设计规范

NB/T 10487-2021

（英文版）

Issued by National Energy Administration of the People's Republic of China
国家能源局　发布
Translation organized by China Renewable Energy Engineering Institute
水电水利规划设计总院　组织翻译
Published by China Water & Power Press
中国水利水电出版社　出版发行
　　Tel: (+ 86 10) 68545888　68545874
　　sales@mwr.gov.cn
　　Account name: China Water & Power Press
　　Address: No.1, Yuyuantan Nanlu, Haidian District, Beijing 100038, China
　　http: //www.waterpub.com.cn
中国水利水电出版社微机排版中心　排版
北京中献拓方科技发展有限公司　印刷
184mm×260mm　16 开本　2.5 印张　79 千字
2024 年 5 月第 1 版　2024 年 5 月第 1 次印刷

Price（定价）：￥410.00

Introduction

This English version is one of China's energy sector standard series in English. Its translation was organized by China Renewable Energy Engineering Institute authorized by National Energy Administration of the People's Republic of China in compliance with relevant procedures and stipulations. This English version was issued by National Energy Administration of the People's Republic of China in Announcement [2023] No. 4 dated May 26, 2023.

This version was translated from the Chinese standard NB/T 10487-2021, *Code for Design of Rare and Endangered Plants and Old and Notable Trees Protection for Hydropower Projects*, published by China Water & Power Press. The copyright is reserved by National Energy Administration of the People's Republic of China. In the event of any discrepancy in the implementation, the Chinese version shall prevail.

Many thanks go to the staff from the relevant standard development organizations and those who have provided generous assistance in the translation and review process.

For further improvement of the English version, any comments and suggestions are welcome and should be addressed to:

China Renewable Energy Engineering Institute
No. 2 Beixiaojie, Liupukang, Xicheng District, Beijing 100120, China
Website: www.creei.cn

Translating organizations:

POWERCHINA Guiyang Engineering Corporation Limited

China Renewable Energy Engineering Institute

Translating staff:

YI Zhongqiang	ZHANG Junwei	YU Weiqi	GAO Guoqing
ZHOU Chao	ZONG Xiao	ZHAO Zaixing	CHEN Fan
LU Bo	WANG Huoyun	CHANG Na	FAN Xinke

Review panel members:

LIU Xiaofen	POWERCHINA Zhongnan Engineering Corporation Limited
JIN Feng	Tsinghua University
QIAO Peng	POWERCHINA Northwest Engineering Corporation

	Limited
QIE Chunsheng	Senior English Translator
ZHANG Ming	Tsinghua University
YAN Wenjun	Army Academy of Armored Forces, PLA
JIA Haibo	POWERCHINA Kunming Engineering Corporation Limited
GAO Yan	POWERCHINA Beijing Engineering Corporation Limited
LI Shisheng	China Renewable Energy Engineering Institute

National Energy Administration of the People's Republic of China

翻译出版说明

本译本为国家能源局委托水电水利规划设计总院按照有关程序和规定，统一组织翻译的能源行业标准英文版系列译本之一。2023年5月26日，国家能源局以2023年第4号公告予以公布。

本译本是根据中国水利水电出版社出版的《水电工程珍稀濒危植物及古树名木保护设计规范》NB/T 10487—2021翻译的，著作权归国家能源局所有。在使用过程中，如出现异议，以中文版为准。

本译本在翻译和审核过程中，本标准编制单位及编制组有关成员给予了积极协助。

为不断提高本译本的质量，欢迎使用者提出意见和建议，并反馈给水电水利规划设计总院。

地址：北京市西城区六铺炕北小街2号
邮编：100120
网址：www.creei.cn

本译本翻译单位：中国电建集团贵阳勘测设计研究院有限公司
　　　　　　　　水电水利规划设计总院

本译本翻译人员：易仲强　张峻玮　喻卫奇　高国庆
　　　　　　　　周　超　纵　霄　赵再兴　陈　凡
　　　　　　　　陆　波　王火云　常　娜　范欣柯

本译本审核人员：

　　刘小芬　中国电建集团中南勘测设计研究院有限公司

　　金　峰　清华大学

　　乔　鹏　中国电建集团西北勘测设计研究院有限公司

　　郄春生　英语高级翻译

　　张　明　清华大学

　　闫文军　中国人民解放军陆军装甲兵学院

　　贾海波　中国电建集团昆明勘测设计研究院有限公司

　　高　燕　中国电建集团北京勘测设计研究院有限公司

　　李仕胜　水电水利规划设计总院

国家能源局

Announcement of National Energy Administration of the People's Republic of China
[2021] No. 1

National Energy Administration of the People's Republic of China has approved and issued 320 energy sector standards including *Code for Integrated Resettlement Design of Hydropower Projects* (Attachment 1), the foreign language version of 113 energy sector standards including *Carbon Steel and Low Alloy Steel for Pressurized Water Reactor Nuclear Power Plants—Part 7: Class 1, 2, 3 Plates* (Attachment 2), and the amendment notification for 5 energy sector standards including *Technical Code for Investigation and Assessment of Aquatic Ecosystem for Hydropower Projects* (Attachment 3).

Attachments: 1. Directory of Sector Standards

2. Directory of Foreign Language Version of Sector Standards

3. Amendment Notification for Sector Standards

National Energy Administration of the People's Republic of China

January 7, 2021

Attachment 1:

Directory of Sector Standards

Serial number	Standard No.	Title	Replaced standard No.	Adopted international standard No.	Approval date	Implementation date
...						
4	NB/T 10487-2021	Code for Design of Rare and Endangered Plants and Old and Notable Trees Protection for Hydropower Projects			2021-01-07	2021-07-01
...						

Foreword

According to the requirements of Document GNKJ [2015] No. 283 issued by National Energy Administration of the People's Republic of China, "Notice on Releasing the Development and Revision Plan of Energy Sector Standards in 2015", and after extensive investigation and research, summarization of practical experience, and wide solicitation of opinions, the drafting group has prepared this code.

The main technical contents of this code include: basic requirements, in-situ conservation design, ex-situ conservation design, small conservation area planning, management and monitoring, and cost estimate.

National Energy Administration of the People's Republic of China is in charge of the administration of this code. China Renewable Energy Engineering Institute has proposed this code and is responsible for its routine management. Energy Sector Standardization Technical Committee on Hydropower Planning, Resettlement and Environmental Protection is responsible for the explanation of specific technical contents. Comments or suggestions in the implementation of this code should be addressed to:

China Renewable Energy Engineering Institute
No. 2 Beixiaojie, Liupukang, Xicheng District, Beijing 100120, China

Chief development organizations:

POWERCHINA Guiyang Engineering Corporation Limited

China Renewable Energy Engineering Institute

Participating development organizations:

POWERCHINA Chengdu Engineering Corporation Limited

POWERCHINA Kunming Engineering Corporation Limited

POWERCHINA Huadong Engineering Corporation Limited

Longtan Hydropower Development Limited Liability Company

China Three Gorges Corporation

Chief drafting staff:

YI Zhongqiang	YU Weiqi	WEI Lang	TANG Zhongbo
ZONG Xiao	ZHANG Junwei	ZHAO Zaixing	JIANG Hao
CHEN Fan	LU Bo	CHANG Li	SUN Yuan

TANG Da	LIU Yuan	DONG Haoping	QIU Liwen
TANG Xing	WU Wenyou	ZHANG Xichuan	ZHAO Jun
ZHOU Chao	WANG Huoyun	LI Ping	ZHANG Yu
MAO Siyu	CHANG Na	WANG Haiwen	ZHAO Zhangguo
CAO Shikai	SUN Rong		

Review panel members:

WAN Wengong	CHEN Guozhu	DAI Xiangrong	JIN Yi
WANG Hailong	ZHANG Rong	LIN Ronggang	DU Sanlin
ZHU Zizheng	LI Min	YANG Long	FANG Sizhao
GUAN Jiajie	SUN Hu	REN Lyu	YANG Taoping
LI Shisheng			

Contents

1	**General Provisions**	1
2	**Terms**	2
3	**Basic Requirements**	3
4	**In-Situ Conservation Design**	4
4.1	General Requirements	4
4.2	Design Data	4
4.3	Protection Scheme	4
4.4	Design Results	6
5	**Ex-Situ Conservation Design**	7
5.1	General Requirements	7
5.2	Design Data	7
5.3	Site Selection	8
5.4	Transplanting Design	8
5.5	Planting Design	9
5.6	Design Results	10
6	**Small Conservation Area Planning**	11
6.1	General Requirements	11
6.2	Planning Data	11
6.3	Planning Tasks and Principle	11
6.4	Site Selection	12
6.5	General Layout	13
6.6	Zone Planning	13
6.7	Supporting Facilities Planning	14
6.8	Design Results	15
7	**Management and Monitoring**	16
7.1	General Requirements	16
7.2	Management	16
7.3	Monitoring	16
7.4	Design Results	17
8	**Cost Estimate**	18
8.1	Preparation Basis and Principle	18
8.2	Cost Estimation Method	18
8.3	Cost Composition	18
8.4	Results	18
Appendix A	**Contents of Design Report on Protection of Rare and Endangered Plants and Old and Notable Trees for Hydropower Projects**	**19**

Appendix B	**Basic Information of Rare and Endangered Plants and Old and Notable Trees for Hydropower Projects** ································	**21**
Appendix C	**Management Record of Rare and Endangered Plants and Old and Notable Trees for Hydropower Projects** ································	**23**
Appendix D	**Monitoring Record of Rare and Endangered Plants and Old and Notable Trees for Hydropower Projects** ································	**24**

Explanation of Wording in This Code······························· **25**
List of Quoted Standards ·· **26**

1　General Provisions

1.0.1　This code is formulated with a view to standardizing the design principles and scope, and unifying the technical requirements for protection of rare and endangered plants and old and notable trees for hydropower projects.

1.0.2　This code is applicable to the design of rare and endangered plants and old and notable trees protection for hydropower projects.

1.0.3　The design of rare and endangered plants and old and notable trees protection for hydropower projects shall meet the overall requirements of regional ecosystem protection, follow the principles of adapting to local conditions, technological feasibility and economic rationality, and coordinate with relevant plans, so as to conserve the biodiversity and the historical and cultural value of old and notable trees.

1.0.4　The design documents on protection of rare and endangered plants and old and notable trees for hydropower projects should be prepared separately, and a special design report and drawings should be provided. Contents of design report on protection of rare and endangered plants and old and notable trees for hydropower projects should comply with Appendix A of this code.

1.0.5　In addition to this code, the design of rare and endangered plants and old and notable trees protection for hydropower projects shall comply with other current relevant standards of China.

2 Terms

2.0.1 in-situ conservation

conservation of rare and endangered plants and old and notable trees as well as their ecosystems and habitats affected by a hydropower project within their original places through avoidance and other measures

2.0.2 ex-situ conservation

conservation of the rare and endangered plants and old and notable trees unavoidably affected by a hydropower project by transplanting them outside their natural habitats

2.0.3 small conservation area

natural habitat area with alternative conservation value, which is selected near the areas affected by a hydropower project, and jointly protected by the project owner and local communities

3 Basic Requirements

3.0.1 The design of rare and endangered plants and old and notable trees protection for a hydropower project shall be based on field investigation on the resources and their habitats, define the protected objects, protection objectives and protection patterns, and propose a design scheme after analysis and demonstration. The field investigation of rare and endangered plants and old and notable trees shall comply with the current sector standard NB/T 10080, *Technical Code for Investigation and Assessment of Terrestrial Ecosystem for Hydropower Projects*.

3.0.2 For the design of rare and endangered plants and old and notable trees protection for a hydropower project, appropriate protection patterns shall be selected against the impacts of the project. Priority shall be given to in-situ conservation for project construction areas, and ex-situ conservation should be adopted for reservoir inundation areas. When a large number of rare and endangered plants are involved in the reservoir inundation areas and ex-situ conservation can hardly achieve the protection objective, a small conservation area shall be established as supplementary.

3.0.3 For the ex-situ conservation design and small conservation area planning for rare and endangered plants, classified protection measures and zoned protection plans shall be formulated according to the habitat conditions in the project area and its surroundings and the biological characteristics of the rare and endangered plants.

3.0.4 The design of rare and endangered plants and old and notable trees protection for a hydropower project shall make full use of the existing conditions. For the protection of each species, a typical design shall be conducted.

3.0.5 For the design of rare and endangered plants and old and notable trees protection for hydropower projects, requirements on information recording and filing of their transplanting process shall be put forward. The basic information of rare and endangered plants and old and notable trees for hydropower projects should be recorded in accordance with Appendix B of this code.

4 In-Situ Conservation Design

4.1 General Requirements

4.1.1 The in-situ conservation design shall define the protected objects and protection areas according to the project construction layout, as well as the range and extent of the impact of reservoir water level fluctuation on the rare and endangered plants and old and notable trees.

4.1.2 The in-situ conservation design shall propose a protection scheme and protection measures, on the basis of the collection and analysis of relevant design information and data of the project.

4.2 Design Data

4.2.1 For the in-situ conservation design, the data on the current status of the rare and endangered plants and old and notable trees affected by the project shall be collected, and the information on the project construction and operation shall be collected as well.

4.2.2 The data on the current status of the rare and endangered plants and old and notable trees shall mainly include:

1. Distribution location, coordinates, altitude, tree orientation, ownership, protection class, and endangerment grade.

2. Species, distribution area, population, plant age, plant height, diameter at breast height or base diameter, crown breadth, and growth status.

3. Habitat stability, existing protection facilities, population of existing communities of the same species in the area and trend of changes, external disturbance factors and extent, external natural environmental conditions, connections with other protected areas, and land ownership.

4.2.3 The information on the project construction and operation shall include construction planning, construction land requisition planning, reservoir impoundment plan and operation schedule.

4.3 Protection Scheme

4.3.1 The in-situ conservation scheme shall be formulated according to the protection requirements, protection conditions and existing protection basis for the rare and endangered plants and old and notable trees.

4.3.2 For in-situ conservation scheme, protection measures shall be selected as required, including fence or wall, warning sign hung or erected, support and reinforcement, and lightning protection.

4.3.3 The design of fences or walls shall meet the following requirements:

  1 The location, type, size and quantity of fences or walls shall be clearly defined in the design, and shall be in harmony with the landscape.

  2 Considering the availability of local materials, the fence should be of the mechanical type made of barbed wire, bamboo or wood, or the green type of barbed shrubs.

  3 The wall should be built with bricks, masonry or concrete. The wall height should be at least 1.2 m above the ground surface, the wall thickness shall be reasonably determined according to the material property and stability requirements, and the wall foundation shall be provided with drains.

  4 An appropriate clearance shall be reserved between the protected objects and the fence or wall.

4.3.4 The design of warning signs hung or erected shall meet the following requirements:

  1 The shape, material, text, font, and hanging or erecting position of signs shall be clearly defined in the design, and shall be in harmony with the landscape.

  2 The hanging position of signs shall be reasonably determined according to the height of the plants, usually 1.2 m to 2.4 m above the ground surface.

  3 The warning sign board should be rectangular or oval, with white text on blue background, and material resisting corrosion and oxidation.

  4 The text on the hung sign shall mainly include the plant species name and Latin name, family and genus, tree age, protection class, conservation values, sign-hanging date, organization and person responsible for management and protection, and complaints hotline.

  5 The text on the erected sign shall mainly include the plant species name and Latin name, family and genus, protection class, conservation values, erecting date of the sign, organization and person responsible for management and protection, and complaint hotlines.

4.3.5 The support and reinforcement design shall meet the following requirements:

  1 Appropriate support and reinforcement patterns shall be determined according to the growth status of the tree and the surrounding environment, and the types and specifications of support and reinforcement materials shall be reasonably determined according to the load on the tree trunks.

2   Severely tilting or unstable trees shall be protected by support, and split trees or trees under fracture risk shall be protected by reinforcement.

3   An elastic cushion shall be set between the tree and the support/reinforcement.

4.3.6   Lightning protection measures shall be taken for isolated and tall trees growing in open areas.

4.4   Design Results

For in-situ conservation design, the following drawings shall be presented:

1   Layout of fences or walls.

2   Typical design drawings of fences or walls.

5 Ex-Situ Conservation Design

5.1 General Requirements

5.1.1 The ex-situ conservation design shall follow the principle of adapting to local conditions, and select appropriate schemes and technical measures according to the ecological characteristics of various rare and endangered plants and old and notable trees.

5.1.2 The ex-situ conservation schemes shall be determined after techno-economic comparison in consideration of factors such as the scale and number of protected objects, the selection of destination, the construction conditions, and the transport conditions.

5.1.3 For the protected objects with a large number and little experience, an ex-situ conservation test of appropriate scale shall be conducted in advance, and the conservation scheme shall be determined accordingly.

5.2 Design Data

5.2.1 For ex-situ conservation design, the data on the current status of the protected objects, the habitat and site conditions of the origin area and the destination, and the transport conditions shall be collected.

5.2.2 The data on the current status of the protected objects shall include those specified in Article 4.2.2 of this code, and those of their main associated and epiphytic plants, the insects that have co-evolution relationships with them, etc.

5.2.3 The data on the habitat and site conditions in both the original area and destination shall mainly include:

1. Information on geographic location, area, ownership, district, etc.
2. Data on topography and geology such as altitude, landform, slope gradient and orientation, groundwater level, and geological disasters.
3. Data on meteorological characteristics such as sunlight, air temperature, accumulated temperature, humidity, precipitation, wind speed, wind direction, evaporation, and meteorological disasters.
4. Soil data such as soil type, physical and chemical properties, and fertility.
5. Vegetation data such as current vegetation types, plant species, community structure, and distribution characteristics.
6. Data on construction conditions such as work space, water sources, and

power supply facilities.

5.2.4 Data on transport conditions shall cover route, haul distance, pavement width, turning radius, loading capacity of bridges and culverts, etc.

5.3 Site Selection

5.3.1 The selection of the destination shall be determined through comparison of alternatives considering the habitat conditions, ownership, construction conditions, cost, impacts on ecosystem, etc.

5.3.2 The destination shall be such that the habitat conditions are highly similar to or better than those in origin areas in terms of altitude, soil type, soil structure, soil fertility, local climate, etc. Besides, the site selection should be coordinated with the construction planning, greening and landscaping of the project.

5.3.3 The land use of the destination shall be reasonably determined based on the type and number of the protected objects and the space requirements for future growth and reproduction. If necessary, multiple sites may be selected.

5.4 Transplanting Design

5.4.1 The determination of the number of the protected objects in ex-situ conservation shall meet the following requirements:

1. The number of transplanted rare and endangered plants in ex-situ conservation shall be reasonably determined considering the ecological protection planning, biological characteristics of rare and endangered plants, biodiversity, etc., which shall be approved by competent authorities.

2. For the old and notable trees that need to be transplanted, the ex-situ conservation scheme shall be developed through special researches, and shall be consented by competent authorities.

5.4.2 The transplanting time shall be determined considering the growth habits of the protected objects, the climatic characteristics of the project area, and the project construction progress.

5.4.3 Before transplanting, biological measurement shall be carried out on the protected objects, and the orientation of the tree shall be marked.

5.4.4 Before tree excavation, the pruning scheme shall be determined based on the balance principle of moisture metabolism, and the root cutting and treatment scheme shall be determined by the tree size.

5.4.5 The plant excavation method shall be reasonably determined considering the growth vigor, the size of root system, local climate, soil characteristics, etc. Old and notable trees should be excavated with soil balls, and herbs and other dwarf plants should be excavated with soil blocks.

5.4.6 The packaging of a plant shall consider its growth characteristics, soil compactness, size of soil ball, haul distance, etc. Old and notable trees should adopt soft packaging with soil balls, or be packed with box board with soil table.

5.4.7 The plant lifting scheme shall be determined considering the weight, space occupation and breast diameter of protected objects, and its protection requirements during lifting, as well as the safety regulations.

5.4.8 The plant transport plan shall be determined in comprehensive consideration of the number, weight, and space occupation of protected objects, the care and protection requirements during transporting, and the transport conditions.

5.5 Planting Design

5.5.1 The planting design shall follow the planning principle of adapting to local terrain, and shall reasonably determine the overall layout based on the functional zoning and the species and biological characteristics of the protected objects.

5.5.2 The design of habitat restoration shall be based on the ecological habitus and original habitat characteristics of the protected objects, and shall provide suitable site conditions and succession space for plant growth by means of earth surface shaping, soil improvement, community reconstruction, etc.

5.5.3 The design of community reconstruction shall determine the reasonable combination of plant species, planting scale and density following the principle of artificial nature and considering the coordination and stability of the inter-species relationship.

5.5.4 The design of site preparation shall determine the terrain treatment plan considering the natural drainage direction of the destination. The site preparation options shall be determined considering the site conditions and plant species.

5.5.5 The planting design shall meet the following requirements:

1. Determine the shape and size of the planting hole according to the distribution of plant root system, the size of the soil ball, and soil conditions.

2. Determine the planting orientation according to the original orientation

of the protected object.

3 Determine the planting depth according to the ecological characteristics and distribution of the plant root system, as well as the terrain, landform, soil texture, groundwater level, and drainage feature of the planting site.

5.5.6 The design of auxiliary measures for the trees to be planted shall put forward the requirements for sun shading, rooting and sprouting, moisturizing and insulation, disaster prevention and lightning protection, support, tagging, etc.

5.6 Design Results

5.6.1 For ex-situ conservation design, the following tables shall be presented:

1 Basic information table of the rare and endangered plants and old and notable trees.

2 Statistical table of the ex-situ conservation test results of the rare and endangered plants.

3 Evaluation table of survival rate of the rare and endangered plants under ex-situ conservation.

5.6.2 For ex-situ conservation design, the following drawings shall be presented:

1 Distribution map of the rare and endangered plants and old and notable trees.

2 Relative location map of the origin area and destination of the transplanted plants and the hydropower project.

3 Destination zoning map.

4 General layout of planting site of the transplanted plants.

5 Vertical layout of planting site of the transplanted plants.

6 Typical planting design drawings.

7 Typical design drawings of excavation, packaging and lifting of the rare and endangered plants and old and notable trees.

6 Small Conservation Area Planning

6.1 General Requirements

6.1.1 The small conservation area planning shall be conducted on the basis of analysis and evaluation of the data collected in the early stage, understanding of the needs of all parties, and clear planning tasks and objectives.

6.1.2 The small conservation area planning shall include the site selection, general layout, zone planning, and supporting facilities planning.

6.2 Planning Data

6.2.1 For the small conservation area planning, the data on hydropower project construction, relevant regional planning, and the status of ecological, natural and social environment shall be collected.

6.2.2 The data on the hydropower project construction shall cover hydropower development pattern, reservoir characteristics and operation scheduling, construction planning, resettlement planning.

6.2.3 The data on relevant regional planning shall cover regional overall development plan, ecological function zoning, ecological construction plan, land use plan, and transportation plan.

6.2.4 The data on ecological environment shall cover the survey, monitoring and evaluation results of regional terrestrial ecosystem and wetland ecosystem, and the data on utilization and protection of ecological resources.

6.2.5 The data on natural environment shall cover regional geology, landform, climate, hydrology, and soil.

6.2.6 The data on social environment shall cover local history, economy, culture, and customs.

6.3 Planning Tasks and Principle

6.3.1 Small conservation area planning shall specify function zoning and supporting facilities through rational site selection and overall layout, to protect biodiversity and to ensure the long-term stable viability, breeding and normal succession of the protected objects.

6.3.2 The planning tasks of a small conservation area shall include:

    1 Reasonably determine the location, scale, range and boundary of the small conservation area.

    2 Define the function zoning and specific planning of the small conservation area, and refine the planning of supporting facilities.

3   Establish a sound management and monitoring system for the small conservation area.

6.3.3   Considering the planning tasks and objectives of small conservation area, the planning principles shall be as follows:

1   Zone reasonably to meet the basic functions of the small conservation area.

2   Give priority to the protection of the existing vegetation and its habitat in the area.

3   Lay out the area in harmony with the surrounding environment and regional features.

4   Give full play to the comprehensive functions of the small conservation area in terms of protection, research, publicity and education.

6.4   Site Selection

6.4.1   According to the protection requirements and patterns, the site of a small conservation area shall be selected after comprehensive comparison and analysis in terms of species distribution, habitat, transport conditions, impacts of economic and social activities, cost, etc.

6.4.2   The land ownership of a small conservation area shall be distinct, and agreed upon by all stakeholders.

6.4.3   The site selection of a small conservation area shall meet the following requirements:

1   Similar habitats and flora communities.

2   Sound habitats, favorable geological conditions, and no major geological hazards or flood influences.

3   Sufficient land area and carrying capacity to support the growth and self-renewal of the protected species.

4   Less disturbance from economic activities.

5   Convenience for the establishment and management of the small conservation area.

6.4.4   The boundary and enclosure of a small conservation area shall be determined considering the growth, breeding, and normal succession of protected plants and peripheral interference.

6.5 General Layout

6.5.1 The general layout of small conservation area shall be determined based on the field investigation and protection condition analysis, considering the various requirements.

6.5.2 The general layout of the small conservation area shall meet the following requirements:

1. Adapt to local conditions and highlight the theme.
2. Reasonably zone the small conservation area by function considering protection conditions, protection objectives and protection requirements.

6.5.3 The small conservation area shall be divided into protected zones and buffer zones, and the division may be readjusted as needed.

6.6 Zone Planning

6.6.1 The zone planning shall specify the function, characteristics and construction items of each zone on the basis of the general layout of the small conservation area.

6.6.2 The planning of protected zones shall meet the following requirements:

1. Strict protection measures shall be taken in the area where protected objects are concentrated.
2. Facilities and activities irrelevant to conservation shall be prohibited in the protected zones.

6.6.3 The planning of buffer zones shall meet the following requirements:

1. The human activities in the buffer zone where protected objects concentrate shall be controlled to lessen disturbance on the protected zone. The buffer zone may be properly developed in some way, but shall not affect the subsistence and breeding of the protected objects or damage the ecosystem.
2. Subzones for transplanting, test and science exhibition shall be set as required to serve the protected zone.

6.6.4 The planning of the transplanting subzone shall meet the following requirements:

1. The transplanting subzone shall provide similar flora community combinations and structural relationship of the natural habitats according to the ecological and biological characteristics of the species transplanted.

2   The transplanting subzone shall provide the land area and space for the subsistence and future breeding and succession of the transplanted plants and their associated plants.

3   The layout shall adapt to the terrain to minimize excavation.

6.6.5   The planning of the test subzone shall meet the following requirements:

1   The area accessible and convenient for observation and test shall be selected according to the characteristics of the test species and the techniques, which may also be used for science exhibition in the future.

2   According to the total number of plants to be transplanted, the minimum number of transplanted plants and the test area shall be determined in the small conservation area.

3   The transplanting test and seedling cultivation test should be arranged on the principle of test before popularization, using advanced and applicable planting technology and research achievements.

4   Favorable conditions such as outdoor nurseries, sun shelters and greenhouses shall be provided for test as required.

6.6.6   The planning of science exhibition subzone shall meet the following requirements:

1   The value, image and characteristics of the small conservation area and the elementary knowledge and legal provisions on the protection of the rare and endangered plants and old and notable trees shall be exhibited.

2   The setup time, scale and characteristics of the small conservation area as well as the protection class, species, quantity and biological characteristics of the main protected objects shall be exhibited.

3   The gate, entrance square and greenbelt shall be set in proper place as the science exhibition area. Landscape nodes shall be planned properly.

6.7   Supporting Facilities Planning

6.7.1   The planning of supporting facilities shall meet the following requirements:

1   The supporting facilities shall be planned to serve the function zones of the small conservation area according to the zone planning.

2   Provided that the basic functions of the small conservation area are satisfied, the planning shall pay more attention to aesthetics, ecotype, and diversity, considering the overall style of the area.

3 The planning shall be forward-looking and satisfy the demand of both short-term construction and long-term development.

6.7.2 The supporting facilities shall include the management building, road system, drainage system, irrigation system, and sign system.

6.7.3 The management building shall be planned according to the area, scale and operation management requirements of the small conservation area.

6.7.4 The road system, including vehicle lanes, service roads, landscape roads and sightseeing walks, shall be arranged considering the existing roads and the local conditions.

6.7.5 The drainage system shall be arranged according to the landform, geology, runoff yield and flow concentration of the small conservation area.

6.7.6 The irrigation system shall be arranged considering the landform, geology, soil, climate, hydrology, and plant growth characteristics. Proper irrigation methods and facility layout shall be selected.

6.7.7 The signs shall be provided at the conspicuous positions of main entrances and exits, intersections and each subzone according to the zone planning of the small conservation area.

6.8 Design Results

For small conservation area planning, the following drawings shall be presented:

1 Map of geographical location of small conservation area.

2 Map of distribution of rare and endangered plants and old and notable trees.

3 General layout of small conservation area.

4 Typical design drawings of small conservation area.

7 Management and Monitoring

7.1 General Requirements

7.1.1 According to different protection options for rare and endangered plants and old and notable trees, a targeted management and monitoring scheme shall be developed, following the principles of rationality, applicability and operability, and adopting advanced technology and methods.

7.1.2 The management and monitoring organization and staff shall be assigned, and the management and monitoring mechanisms, forms and rules shall be proposed, according to the ownership, distribution range and survival difficulty of the rare and endangered plants and old and notable trees and the small conservation area.

7.2 Management

7.2.1 The management for in-situ conservation shall include the care of the protected objects and the maintenance of the protection measures.

7.2.2 The management for ex-situ conservation shall include post-planting care and long-term management of protected objects. Management approaches including hanging tags, supporting, shading, watering, pruning, lightning protection, freezing protection, pest control, and disaster prevention and control should be adopted.

7.2.3 The management for the small conservation area shall be conducted considering the management requirements and items of in-situ conservation and ex-situ conservation.

7.2.4 The management design shall define the requirements for registering the basic information about the rare and endangered plants and old and notable trees, as well as recording the management activities. Management record of rare and endangered plants and old and notable trees for hydropower projects should comply with Appendix C of this code.

7.3 Monitoring

7.3.1 Specific monitoring requirements shall be put forward according to the ecological habitus and environmental factors of the protected objects. The monitoring items, methods, time period, frequency, etc. shall be defined. The monitoring data of rare and endangered plants and old and notable trees shall be recorded. Monitoring record of rare and endangered plants and old and notable trees for hydropower projects should comply with Appendix D of this code.

7.3.2 The monitoring shall cover the protected objects and environmental factors, with emphasis on the transplanted plants.

7.3.3 The monitoring time and frequency should be relatively fixed. The protected objects susceptible to peripheral interference shall be monitored more frequently.

7.3.4 The monitoring organization should be responsible for regularly analyzing and evaluating the monitoring data and mastering the changes of the protected objects and their habitats.

7.4　Design Results

The management and monitoring design shall put forward a schedule for monitoring the rare and endangered plants and old and notable trees for hydropower projects.

8 Cost Estimate

8.1 Preparation Basis and Principle

8.1.1 The cost estimate preparation of the rare and endangered plants and old and notable trees protection works shall comply with the current sector standard NB/T 35033, *Preparation Regulation for Special Investment on Environmental Protection Design of Hydropower Project*.

8.1.2 The price level used in the cost estimation of the rare and endangered plants and old and notable trees protection works shall be consistent with that of the main works.

8.2 Cost Estimation Method

The cost estimation method for the rare and endangered plants and old and notable trees protection works shall comply with the current cost estimation specifications and standard rates for hydropower projects.

8.3 Cost Composition

8.3.1 The cost of the rare and endangered plants and old and notable trees conservation works shall include the cost of the construction works, the equipment and installation works, the special research, the management and monitoring, etc.

8.3.2 The independent cost shall be composed of the project construction management cost, the scientific research, survey and design cost, and taxes.

8.3.3 The basic contingency shall be calculated as a percentage of the sum of the cost of the conservation works and the independent cost. The percentage for the basic contingency shall be determined based on the difficulty level of the protection works.

8.3.4 The initial-stage management and monitoring cost of the protected objects shall be included in the cost of rare and endangered plants and old and notable trees protection. The subsequent cost shall be borne by the owner of the protected objects.

8.4 Results

For the cost estimation for protection of the rare and endangered plants and old and notable trees, the following tables shall be presented:

1 Table of total cost.

2 Table of cost breakdown.

3 Table of annual cost.

Appendix A Contents of Design Report on Protection of Rare and Endangered Plants and Old and Notable Trees for Hydropower Projects

Foreword

1 Introduction

1.1 Background

1.2 Preparation Purpose

1.3 Scope

1.4 Preparation Basis

1.5 Design Philosophy and Principle

1.6 Work Process

2 Project Overview

2.1 Overview of River Basin

2.2 Overview of Project

2.3 Natural Environment Status

2.4 Environmental Impacts and Countermeasures

3 Status of and Impacts on Rare and Endangered Plants and Old and Notable Trees

3.1 Species

3.2 Basic Characteristics

3.3 Current Distribution

3.4 Habitat Conditions

3.5 Impact Analysis

4 Overall Protection Scheme

4.1 Protection Tasks and Objectives

4.2 Protection Scheme

4.3 General Layout of Protection Measures

5 Protection Design

NB/T 10487-2021

5.1 In-Situ Conservation

5.2 Ex-Situ Conservation

5.3 Small Conservation Area Planning

6 Management and Monitoring

6.1 Management

6.2 Monitoring

7 Cost Estimate

7.1 Preparation Basis and Principles

7.2 Cost Estimation Method

7.3 Cost Composition

8 Conclusions and Suggestions

8.1 Conclusions

8.2 Suggestions

Attachments:

1 Design Contract/Agreement

2 Approval of the Project Environmental Impact Assessment Documents

Attached Tables:

1 Basic Information of the Rare and Endangered Plants and Old and Notable Trees

2 Tables of Total Cost and Cost Breakdown

Attached Figures:

1 Distribution of Rare and Endangered Plants

2 Distribution of Old and Notable Trees

3 General Layout of Protection Measures

4 Layout of Protection Measures

5 Typical Design Drawings of Protection Measures

6 Layout of Small Conservation Areas

7 Layout of Monitoring Sites

Appendix B Basic Information of Rare and Endangered Plants and Old and Notable Trees for Hydropower Projects

Table B Basic information of rare and endangered plants and old and notable trees for hydropower projects

No.				
Chinese name			Family	
Latin name			Genus	
Origin area	Location relative to the hydropower project			
	Longitude and latitude		Altitude	
Distribution characteristics	○Dispersed ○Clustered			
Biological characteristics	Plant height		DBH/base diameter	
	Plant age		Crown diameter	
	Life form	○Arbor ○Shrub ○Vine ○Herb ○Others		
	Growth vigor	○Vigorous ○Ordinary ○Poor ○Dying ○Dead		
	Modes of reproduction	○Sexual reproduction ○Asexual reproduction		
	Phenological period	○Seedling stage ○Vegetative period ○Flowering phase ○Fruiting season ○Post-fruiting period		
Ecological characteristics				
Protection class				
Site conditions	Slope gradient		Slope direction	
	Slope position		Soil type	
	Soil thickness		pH	
	Climatic characteristics		Community type	
Protection scheme and measures	In-situ conservation	Protection measures		

Table B *(continued)*

Protection scheme and measures	Ex-situ conservation	Location relationship between the destination and the hydropower project	
		Geographical coordinates and altitude of the destination	
		Habitat conditions of the destination	
		Ex-situ conservation time	
		Conservation measures	
	Small conservation area	Location relationship between the small conservation area and the hydropower project	
		Geographical coordinates and altitude of the small conservation area	
		Habitat conditions of the small conservation area	
		Time of transplanting to the small conservation area	
		Protection measures	
Description of key features of protected objects			
Records of history, culture and legend			
Photo No.		Photographed by	Date
Remarks			
Recorded by		Recording time	

NOTE Marking √ at the ○ denotes the chosen item.

Appendix C Management Record of Rare and Endangered Plants and Old and Notable Trees for Hydropower Projects

Table C Management record of rare and endangered plants and old and notable trees for hydropower projects

No.				
Chinese name		Family		
Latin name		Genus		
Protection approach		Protection time		
Site conditions				
Management measures				
Growth conditions before management				
Growth conditions after management				
Management effects				
Description of anomalies				
Photo No.		Photographed by		Date	
Remarks				
Management personnel		Management time		

Appendix D Monitoring Record of Rare and Endangered Plants and Old and Notable Trees for Hydropower Projects

Table D Monitoring record of rare and endangered plants and old and notable trees for hydropower projects

No.				
Chinese name		Family		
Latin name		Genus		
Climate		Temperature		
Growth status of plant	Plant height		DBH/base diameter	
	Plant age		Crown diameter	
	Growth vigor	○Vigorous　○Ordinary　○Poor　○Dying　○Dead		
Site conditions	Soil conditions			
	Plant diseases and insect pests			
	Natural disasters			
Protection facilities				
Other issues				
Photo No.		Photographed by	Date	
Remarks				
Monitored by		Monitoring time		

NOTE　Marking √ at the ○ denotes the chosen item.

Explanation of Wording in This Code

1   Words used for different degrees of strictness are explained as follows in order to mark the differences in executing the requirements in this code.

   1) Words denoting a very strict or mandatory requirement:

   "Must" is used for affirmation; "must not" for negation.

   2) Words denoting a strict requirement under normal condition:

   "Shall" is used for affirmation; "shall not" for negation.

   3) Words denoting a permission of a slight choice or in an indication of the most suitable choice when conditions permit:

   "Should" is used for affirmation; "should not" for negation.

   4) "May" is used to express the option available, sometimes with the conditional permit.

2   "Shall meet the requirements of…" or "shall comply with…" is used in this code to indicate that it is necessary to comply with the requirements stipulated in other relative standards and codes.

List of Quoted Standards

NB/T 10080, *Technical Code for Investigation and Assessment of Terrestrial Ecosystem for Hydropower Projects*

NB/T 35033, *Preparation Regulation for Special Investment on Environmental Protection Design of Hydropower Project*